D1429260

John Lewis-Stempel's books include *Meadowland* and *Where Poppies Blow*, both of which have won the Wainwright Prize for Nature Writing. His other books include the *Sunday Times* bestsellers *The Running Hare* and *The Wood*, both BBC Radio 4 Book of the Week. He is a Magazine Columnist of the Year for his notes on nature in *Country Life*.

www.penguin.co.uk

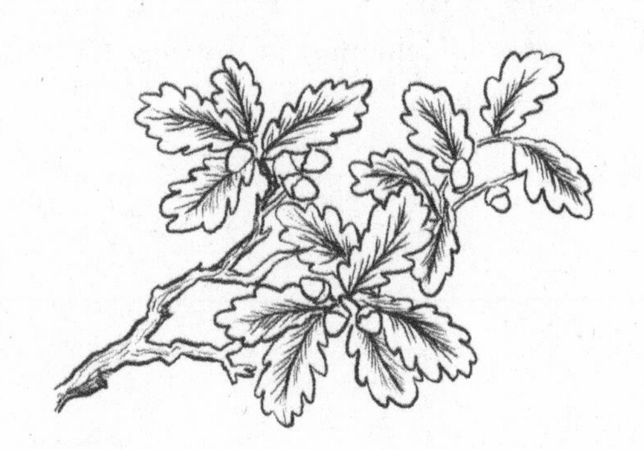

The Glorious Life of the OAK

John Lewis-Stempel

Doubleday

LONDON · NEW YORK · TORONTO · SYDNEY · AUCKLAND

TRANSWORLD PUBLISHERS
61–63 Uxbridge Road, London W5 5SA
www.penguin.co.uk

Transworld is part of the Penguin Random House
group of companies whose addresses can be found at
global.penguinrandomhouse.com

First published in Great Britain in 2018 by Doubleday
an imprint of Transworld Publishers

A CIP catalogue record for this book
is available from the British Library.

ISBN 9780857525819

Typeset in 11/14.5pt Goudy Oldstyle Std India
by Integra Software Services Pvt. Ltd, Pondicherry

Printed and bound in Great Britain by Clays Ltd, Elcograf S.p.A.

Penguin Random House is committed to a sustainable future for
our business, our readers and our planet. This book is made from
Forest Stewardship Council® certified paper.

5 7 9 10 8 6 4

The Round Oak

The Apple top't oak in the old narrow lane
And the hedge row of bramble and thorn
Will ne'er throw their green on my visions again
As they did on that sweet dewy morn
When I went for spring pooteys and birds nest to look
Down the border of bushes ayont the fair spring
I gathered the palm grass close to the brook
And heard the sweet birds in thorn bushes sing

I gathered flat gravel stones up in the shallows
To make ducks and drakes when I got to a pond
The reed sparrows nest it was close to the sallows
And the wrens in a thorn bush a little beyond
And there did the stickleback shoot through the pebbles
As the bow shoots the arrow quick darting unseen
Till it came to the shallows where the water scarce drebbles
Then back dart again to the spring head of green

The nest of the magpie in the low bush of white thorn
And the carrion crows nest on the tree o'er the spring
I saw it in March on many a cold morn

When the arum it bloomed like a beautiful thing
And the apple top't oak aye as round as a table
That grew just above on the bank by the spring
Where every Saturday noon I was able
To spend half a day and hear the birds sing

But now there's no holidays left to my choice
That can bring time to sit in thy pleasures again
Thy limpid brook flows and thy waters rejoice
And I long for that tree – but my wishes are vain
All that's left to me now I find in my dreams
For fate in my fortune's left nothing the same
Sweet Apple top't oak that grew by the stream
I loved thy shade once now I love but thy name

John Clare (1793–1864)

Contents

Prologue: A Meeting with an Oak at Midnight 1

Introduction: From Little Acorns … 7

I: Heart of Oak: The Oak in British History 15

II: Oakwatch: The Life Cycle of the Oak 33

III: 'To Assuage Inflammations':
The Oak as Medicine, Food and Drink 57

IV: Britain's Mightiest Oaks 65

V: The Thunder Tree: Oak in Folklore, Myth and Legend 77

Timeline 86

Picture Credits 88

PROLOGUE

A Meeting with an Oak at Midnight

WE ALL BLAMED each other, but someone had left the back door open, letting the Border terrier escape. It was after eleven, just as we were going to bed. No one wants a dog running free at night, especially a dog who has form for barking at sheep.

I had a hunch the dog had headed for the wood down below the house, into foxland. A reynard is a *bête rouge* to a Border terrier. So I grabbed a labrador to trail the Border terrier trailing the fox. Dog follow dog follow dog.

We went down the yard, through the garden, into the winter wood, the Labrador and I panting white in the night. Looping around in my head was our family's signature tune: the Baha Men's 'Who Let The Dogs Out?'

Who? Who?

I didn't need the torch I'd taken: the stars over the hills of West Herefordshire were light enough. We took the pale line of the path, swerving the iron trees, into the plots of starshine. Black and white, black and white.

Halfway into the wood I could hear rapid-fire terrier-type barking, but with a curious bassoon boom.

The foxes' den was under a pile of scrap metal bulldozed into the wood by a farmer decades ago. Our foxes are the wombling foxes, making good use of the rubbish they find.

The Border terrier had his head in the foxes' corrugated iron hallway so his bark was echoed and amplified. Tin-can speakers.

On the call 'Rupert! Here!' he came willingly enough, a sense of duty done. The panic over, the dogs and I started for home and bed. It was while we were walking back that I realized the gift the Border terrier had given me.

It is not easy to tear oneself from the seduction of civilization; a nightwalk in a cold wood does not necessarily appeal.

So, if it had not been for Rupert running off I'd not have seen the branches of the birch hanging with stars, or heard the barn owl's scream stone-skipping over the frozen fields.

Or met the trees.

I must have walked past the oak, ash and beech on the bend a thousand times, but that midnight in December I saw them differently.

A tree at night is a tree in the wild, located in its own space and time. Outside the desire of our feeble pinky-grey cells to arrange the world in an ordered scheme it doesn't really have. Outside pure reason.

Never before that minute had I understood – really understood – that the reading of trees depends on their type, time of day, time of year. In winter, in solitude and ice-star skies, the beech was a vaunting column to support the universe; the ash was sulky and untouchable, the middle child; and the oak?

The oak was the mighty giant, who held the ball moon in goblet fingers.

Contemporary arborists like to portray trees as wholly sentient beings, like Ents, the tree people, in Tolkien's *Lord of the Rings*. Ents talk, fall in love.

I'm not a fan of bigging up trees as Ents. To suggest trees are as sentient as primates is to diminish them, and us.

Yet, standing under the old oak, it was difficult not to invest it with personhood. I touched its bark, which was wrinkled, warm and dry; old person's skin. But inside, the oak's heart was solid, dependable, which is the nature of oaks.

I turned on the flashlight: green-man faces peered at me from the trunk. Then, I walked around the oak, through the cracked paving of light and shade, a circle of five strides, and understood the tree in another way: the oak is the wooden tie between heaven and earth, the linchpin of the British landscape.

It occurred to me then that oaks had always been there in the landscape of my memory, unassuming, only half glimpsed, but fixtures nonetheless: the hollow oak at Abbeydore which was the kids' den, and the wild bees' home; the bellied-up oak at Lugwardine I jumped Owen the pony over; the panelling in the hall at school; the leaning oak by the Escley that produced acorns by the bucketload for Gloucester Old Spot pigs; that comfortless pew in Hereford Cathedral where I sat for Confirmation; the oak by the village bus stop that provided the acorn 'bullets' for the toy shotgun my father gave me for my eleventh Christmas.

The oak on the bend was over two hundred years old, so was a sapling when Nelson, in ships of oak, bested the French fleet at Trafalgar. Looking up at the oak in the star-sky, I wondered what events this tree in remote Herefordshire had witnessed. Country dramas, I supposed, such as the rutting of the deer, the birth of the birds, the death of the hares. Small things, but great in their way.

No, a tree is not an Ent. But of all the trees, it is the oak which speaks to us.

INTRODUCTION

From Little Acorns ...

The oak fattens the flesh of swine
for the children of men. Often it traverses the gannet's
bath, and the ocean proves whether
the oak keeps faith in honourable fashion.

Old English rune poem, C.AD 900

IT IS THE conviction of the British that the oak is special to them, them alone. This will be news to the United States whose Congress passed legislation in 2004 naming the oak as America's national tree. Or indeed to the Baltic states of Latvia, Lithuania and Estonia, all of whom claim the oak as their patriotic arboreal emblem.

If, as some say, the oak defines the British countryside, it distinguishes foreign ground too. The 'bocage' of south-west France is field-fenced with oaks, so that in autumn, when the leaves turn, the region has the appearance of thousands of gold frames laid on the ground. The oak has roots stretching back to all the old major European cultures. The oak was sacred to Zeus, whose will was made manifest by the rustling of the tree's leaves. The first British appearance of the proverb 'from little acorns mighty oaks grow' comes in Chaucer's *Troilus and Criseyde*, c.1385:

> *Or as an ook comth of a litel spir,*
> *So thorugh this lettre, which that she hym sente,*
> *Encressen gan desir, of which he brente.*

Yet the proverb was of ancient Roman origin. In the first century AD the Roman scholar Pliny the Elder explained the sacredness of oak to the Gauls, whose heartland was France. Saint Boniface (680–754) had Donar's Oak, revered by the pagans, destroyed. This was at Geismar, south Germany.

And yet . . . The oak does indeed have a privileged place in the heart of the British, their landscape and their history. Sorry, everyone else. There is a simple statistical proof of this exceptional relationship: Britain has more ancient oaks than all other European countries combined. If, for example, all the oaks with a girth (circumference of trunk) greater than 9.00m are totted up, there are 115 in England and only 96 in all the rest of Europe. More than half the ancient oaks in the world are in Britain, those over four hundred years old. There are 3400 of them. The oak is the most common tree species in the UK.

The oak, as the Anglo-Saxon rune poem attests, is the quintessential British tree, the one that everyone recognizes, with its massive domed appearance, its acorns, its artistically lobed leaves.

The majority ethnic British – the Angles, the Saxons, the Norse – came to these isles across the North Sea ('the gannet's bath') in longships made of oak. For centuries, the oak touched – literally – every part of a Briton's life, from cradle to coffin. The oak spanned the house in which he or she lived, provided the acorns to feed the pigs. It was oak that made the 'wooden walls' of Nelson's navy, and it was the navy that allowed Britain to rule the world. Oak was charcoal for smelting in the first foundries of the Industrial Revolution, the axle for the haywain, tannin for leather, shelter for livestock, shade for the courting couple. Ink from oak galls recorded the details of life in the parish registers, the course of events in the court annals. An oblong of oak was the raw material for a boy's first woodwork lesson ('A Simple Coat Peg, With Mortice and Tenon Joint'), while the roots of the tree were the imagined home of gnomes in the children's classic *The Little Grey Men* by BB. Or were for this boy in the 1970s, at least.

Even in the digital Apple age, a real oak has resonance. Say the word, and notions of fortitude, antiquity, pastoralism swell up.

The Druids worshipped the oak as a semi-sentient being. Stand before a venerable oak, as I did on that December night years ago, and the humanness of the tree is difficult to resist, even to those sworn to anti-idolatry. John Aubrey, the seventeenth-century 'scientific' antiquary, wrote: 'When an oak is felling, before it falls it gives off a kind of shrikes or groanes that may be heard a mile off, as if it were the genus of oak lamenting.' When the diarist and vicar of Clyro, Francis Kilvert, visited the oaks of Moccas Park in 1876, he described them memorably as 'grey, gnarled, low-browed, knock-kneed, bowed, bent, huge, strange, long-armed, deformed, hunch-backed, mis-shapen oak "men"'. I saw the oaks of Moccas last week; although they have leprotically lost limbs, the grey men march on.

In Alfred Tennyson's poem 'Talking Oak' (1837–8) the narrator confides so much in the old oak tree that the tree replies, and tells him that the woman he loves is superior to all the other young women he has seen in his five hundred years of life. When J. R. R. Tolkien created Ents, the walking tree beings, it was the oak which inspired him.

An old oak is a character, which is why so many are named, and even visited by admirers, celebrated in song, portrayed in legend.

Actually, there are two oaks native to Britain. The English, common or pendunculate oak (*Quercus robur*), and the durmast or sessile oak (*Quercus petraea*). The trees are similar in appearance and grow to roughly the same height (20m to 25m average). They can be differentiated so: the pendunculate bears its acorns on long stalks – elves use them for their tobacco pipes in fairy stories – while its leaves have little lobes at the base. The sessile has a short stalk for the acorn and a long stalk for the leaf, which lacks lobes.

That said, a British oak is a British oak. The quality of the timber is virtually identical, as is their shape, and their fruit. Anyway, they hybridize madly.

Also both *robur* and *petraea* share the quality that causes the contemporary reverence for the oak. No other tree in the forest, the park, the field provides such a haven for nature. A mature oak is a universe in itself, home to upwards of a thousand species.

John Evelyn, the diarist who might be said to be our first woodsman, wrote in his celebrated *Sylva* (1664) with reference to the oak: 'As long as the Lion holds his place as king of beasts, and the Eagle as king of birds, the sovereignty of British trees must remain to the oak.'

It must.

Oak Tree

I took an acorn and put it in a pot.
I then covered it with earth, not a lot.
Great pleasure was mine watching it grow.
The first budding green came ever so slow.
I watered my plant twice a week,
I knew I would transplant it down by the creek.
One day it will be a giant oak,
To shield me from the sun a sheltering cloak.
Lovers will carve their initials in the bark,
An arrow through a heart they will leave their mark.
It will shelter those caught in a fine summer's rain,
Under its leafy bows joy will be again.
Creatures of the wilds will claim it for their own,
Squirrels will reside here in their own home.
Birds will build nests and raise their young,
They will sing melodies a chorus well sung.
Under its branches grass will grow,
Here and there a wild flower its head will show.
My oak tree for hundreds of years will live.
Perhaps the most important thing I had to give.

George Bernard Shaw (1856–1950)

I

Heart of Oak: The Oak in British History

'This noble tree, the monarch of the wood, the boast and bulwark of the British nation.'

William Boutcher, eighteenth-century landscape designer and nurseryman

IN THE BEGINNING was the wildwood.

When the last Ice Age ended, the British Isles again became suitable for vegetation. Trees which had retreated during the glaciation came north again. The first to colonize the tundra were birch, aspen and sallow. These were followed by pine and by hazel; then alder and, *c.*9000 years ago, oak. (Lime, elm, holly, ash, beech, hornbeam and maple were the arboreal Johnny-come-latelies.) For nearly four millennia Britain was all but covered, from silver sea to silver sea, by 'mixed oak forest'. Then came the Neolithic farmers who felled swathes of the wildwood, the regrowth eaten by their browsing cattle and sheep. Half of Britain ceased to be wilderness by 500 BC, the Iron Age. All who came cut down the trees; the Celts, the Romans, but above all the Anglo-Saxons, who were farmers to the tips of their toiling fingers. By the time of the Domesday Book, the great Norman survey of 1086, England was not very wooded. According to the landscape historian Oliver Rackham, out of 12,580 settlements for which adequate information is given, only 6208 possessed wood.

In that scant woodland, it was the oak which mattered. A running theme of Domesday is 'pannage', the right or privilege of feeding swine in woodlands, from fallen tree fruits, predominantly acorns and beech nuts. (Pannage is from the Latin *pastionaticum* via the Old French *pasnage*, 'pasturing'.) As often as not, the woods and forests of Domesday are valued mainly for the mast they produced, thus the number of hogs they could fatten. The entry for Midleton, in Surrey, is typical: 'the wood yields forty hogs for pannage, and eleven for herbage'. For these rights of pannage the lord of the manor received a rent or payment, one pig in ten or a fee ranging from a halfpenny to four pence a pig. An additional benefit for the landowner was that the porcine's scoffing of the acorns, which contain toxic tannic acid, thus removed a source of poison to cattle and horses, which also shared the woodland. Edmund Spenser's *The Shepheardes Calender* (1579) emphasizes the importance of pannage in Elizabethan times:

> *There grew an aged Tree on the greene,*
> *A goodly Oake sometime had it bene,*
> *With armes full strong and largely displayed* [...]
> *Whilome had been the King of the field*
> *And mochell mast to the husband did yielde,*
> *And with his nuts larded many swine.*

The word acorn is interestingly derived. *AEcern* in Old English has the same stem as *acer*, a field, suggesting that the acorn was regarded as mast, as field food for pigs, back in the Dark Ages. Popular etymology turned it into 'acorn' or 'corn' of the oak. Pannage was merely a Norman term for a pre-existing practice.

By the twentieth century the practice of pannage had all but died out, with the exception of in the New Forest, where the Forestry Commission still grants up to sixty days a year (occasionally more in a glut) for commoners to exercise their 'Common of Mast'. About 200–600 pigs now enjoy this amenity, compared to 6000 in the nineteenth century.

What did the Normans ever do for us? Provided the conditions for oaks to grow to old age. The Normans took over *la toute Angleterre* in 1066, meaning all the land therein belonged to William the Conqueror. Whole tracts were reserved by the King and his barons for deer-hunting. It is in these deer parks of Norman origin or facsimile that about 50 per cent of England's ancient oaks can be found today.

Hunting deer required two things – relatively open woodland (to allow hunters to view the deer they were hunting) and deer, lots of deer. The sheer number of oak-shoot-browsing deer together with the acorn-consuming swine kept woodland open. Oaks are slow-growing and as a result can easily become overshadowed, thus killed, in a crowded forest. Deer

parks were ideal habitats for oaks to become truly ancient in – and that is what, courtesy of William the Conqueror and his nobles, happened in England. The single greatest concentration of ancient oaks is in a deer park at Blenheim Palace estate in Oxfordshire, which has 112 ancient oaks. Among these is Britain's oldest oak, which started growing in AD 970.

The oak woods of old England had further purpose, beyond sports ground and pig trough. Oak was the peerless tree for building timber. Oaks produce one of the hardest and most durable timbers; indeed, *robur* is the Latin for 'strength'. The wood from the oak tree provided the frame for houses, the supports for the roofs of great buildings. A typical Elizabethan farmhouse used 330 oak trees in its construction; 700

oaks supported Norwich Cathedral's roof. Britain's most complex wooden structure, the fourteenth-century roof of Westminster Hall, was preconstructed from 600 oak trees, assembled on site. Geoffrey Chaucer (1343–1400) called it 'byldere oak', and Edmund Spenser concurred in the *Faerie Queene*, c.1589, 'The builder-Oake, sole king of forests all'.

Oak timber shrinks and braces as it seasons, meaning it gets tougher yet as it ages. I can testify to the enduring structural properties of oak: we once lived in a seventeenth-century farmhouse; the stone walls crumbled and decomposed, but the oak frame never bowed. Oak timber can stand a thousand years.

As well as being the prime timber, oak was the most expensive timber; the poor huddled masses lived in houses and hovels with frames of elm, or aspen. Oak panelling was the discreet signifier of wealth and aspiration; the Green Dragon Hotel in Hereford, the place to stay locally for a century, had its dining room panelled with oak sliced from a single tree felled on a nearby farm.

Oak built more than dwellings and places of worship. In the heyday of oaken timber, the tree provided the planks, keels and ribs of Britain's warships, and the merchant service. A signal advantage of the oak in shipbuilding is that some branches are naturally crooked; grown bowed timber with an unbroken grain along the curve is stronger than timber cut to a curve and simplifies the construction of durable hulls. A man-of-war required the timber of three thousand mature oaks. In 1580 Queen Elizabeth, acutely aware of the need for quality timber for the navy, assented to Lord Burghley's order to 'empale' 1 acre of Cranbourne Walk, in Windsor Great Park, and sow it with acorns – the first record of a deliberate oak plantation.

The oaks of England, sawn and shaped into the floating gun-platforms of HMS *Victory* and her kind, were Britain's defence against Napoleonic invasion. David Garrick's 'Heart of Oak', which became the Royal Navy's marching song, is as rousing an anthem as the Georgian age could compose, and testament in choral song to the military efficacy of English blood and oak wood combined:

Heart of oak are our ships, jolly tars are our men,
We always are ready; steady, boys, steady!
We'll fight and we'll conquer, again and again.

Oak mania swept the nation. If all that stood between Britain and perfidious France were the wooden walls of the navy, then oaks, and more oaks, were needed. Admiral Lord Collingwood, one of Nelson's 'band of brothers', walked the lanes of Northumberland with a pocketful of acorns to plant in hedgerows so that England's ships should not want for oak trees. Invited to the homes of the great and good, he walked their broad acres surreptitiously letting acorns fall from an intentional hole in the pocket of his breeches.

Other landowners positively vied with Royal Navy officers as to who could plant the most acorns on behalf of British seapower and exploration. Captain James Cook (1728–79) sailed the blue planet in

HMS *Endeavour*, a ship with a hull of white oak, copper-sheathed to protect against shipworm.

On the road between Coventry and Kenilworth, a Mr Gregory, of Stivichall, planted a mile-long avenue of oaks, three deep on each side of the route. For this patriotic, public-spirited planting on behalf of the senior service's future needs, Mr Gregory was granted the right to append 'supporters' to his coat of arms, a distinction usually only allowed the peerage and knightage.

The all-time champion acorn-planter was the Lord Lieutenant of Cardiganshire, Colonel Thomas Johnes, who between 1795 and 1801 planted 922,000 oaks.

Such endeavours were all the more necessary because of the demands of nascent industry. One ironmaster is said to have near-eradicated the oaks of Cannock Chase between 1550 and 1580. The Enclosure Acts, by which common land was privatized for a more intensive agriculture, also grubbed up oaks galore. John Clare, the peasant poet, was eloquent on the destruction of the oaks that had populated his childhood:

Summer pleasures they are gone like to visions every one
And the cloudy days of autumn and of winter cometh on

I tried to call them back but unbidden they are gone
Far away from heart and eye and for ever far away
Dear heart and can it be that such raptures meet decay
I thought them all eternal when by Langley Bush I lay
I thought them joys eternal when I used to shout and play
On its bank at 'clink and bandy' 'chock' and 'taw' and
ducking stone
Where silence sitteth now on the wild heath as her own
Like a ruin of the past all alone
[. . .]

When jumping time away on old cross berry way
And eating awes like sugar plumbs ere they had lost
the may
And skipping like a leveret before the peep of day
On the rolly polly up and downs of pleasant swordy well
When in round oaks narrow lane as the south got black
again
We sought the hollow ash that was shelter from the rain
With our pockets full of peas we had stolen from the
grain
How delicious was the dinner time on such a showry day
O words are poor receipts for what time hath stole away
The ancient pulpit trees and the play
[. . .]

By Langley Bush I roam but the bush hath left its hill
On cowper green I stray tis a desert strange and chill
And spreading lea close oak ere decay had penned its will
To the axe of the spoiler and self interest fell a prey
And cross berry way and old round oaks narrow lane
With its hollow trees like pulpits I shall never see again
Inclosure like a Buonaparte let not a thing remain
It levelled every bush and tree and levelled every hill
And hung the moles for traitors – though the brook is running still
It runs a naked brook cold and chill [. . .]

The loss in oak numbers, but not of 'characters', was to an extent made up by trees planted in the new hedges of eighteenth-century enclosure, as well as the acorns for the navy.

The oak came to symbolize national liberty. It's there in Clare's poem, where the oak stands against the tyranny of Bonaparte and the tyranny of 'Inclosure'. The MP and man of letters Horace Walpole claimed that an olden oak was a symbol of freedom, since in a country ruled by a despot it would be appropriated for its worth as timber, and Romantic histories told that British liberty had begun, not with Magna Carta, but in the oak wood with the Druids and their rites. The tree represented the ancient liberties, before the 'Norman yoke'.

The monarchy, ever astute when it came to its own survival, dressed itself in legends and connotations of oak. After all, oaks were stately, long-lived, immovable fixtures on the landscape. Shireoaks is the name of the village where Nottinghamshire, Yorkshire and Derbyshire converge. So-called Gospel Oaks marked parish boundaries, under which the gospel was read at Rogation-tide, when the bounds were beaten. So, in the lines of Robert Herrick (1591–1674):

Dearest, bury me
Under that holy-oke, or gospel-tree;
Where, though thou see'st not, thou may'st think upon
Me, when you yeerly go'st procesion.

By associating with the oak, monarchs hoped the tree's majestic qualities would rub off on them. King Edward the Confessor took an oath under a large oak in Highgate, London, to keep and defend the laws of England. Royal business was conducted under oaks; Edward I is said to have called a meeting of parliament in Sherwood Forest in 1290 on his way to hammer the Scots. English kings made a habit of hiding in oaks, a handy symbolic gesture; Henry VI hid in the

King's Oak at Irton Hall, Holmrook, Cumbria, after the battle of Towton in 1461 during the Wars of the Roses. However, the oak as a symbol of the British monarchy really took off during the English Civil War, by virtue of Charles II's stay in the Boscobel Oak after losing the battle of Worcester in 1651.

With the Restoration in 1660, the King's birthday of 29 May became Royal Oak Day, Oak Apple Day, Oak-ball Day, or Shick-shack Day, when a piece of oak (the 'shick-shack') was worn on the person or placed on the house. Oak Apple Day was declared a public holiday by parliament, and its official celebration continued until 1859. The date is still celebrated as Founder's Day at the Royal Hospital Chelsea.

The Royal Oak is the third most common pub name in Britain. Since the Restoration there have been at least eight warships called HMS *Royal Oak*.

Concrete and steel killed the oak as the prime material for British buildings and ships. The tree has less than a decimal of its old usefulness and ubiquity. Time, though, has done nothing to lessen the place of the oak in the hearts and minds of the British. The National Trust, the charitable curators of the nation's heritage, have as their symbol a sprig of oak leaves and an acorn.

Naturally.

The Brave Old Oak: A Traditional English Folk Song

A song to the oak, the brave old oak,
Who hath ruled in the greenwood long;
Here's health and renown to his broad green crown,
And his fifty arms so strong.
There's fear in his frown when the sun goes down,
And the fire in the west fades out;
And he showeth his might on a wild midnight,
When the storm through his branches shout.

Then here's to the oak, the brave old oak,
Who stands in his pride alone;
And still flourish he, a hale green tree,
When a hundred years are gone!

In the days of old, when the spring with cold
Had brightened his branches gray,
Through the grass at his feet crept maidens sweet,
To gather the dew of May.
And on that day to the rebeck gay
They frolicked with lovesome swains;
They are gone, they are dead, in the churchyard laid,
But the tree it still remains.

Then here's, *etc.*

He saw the rare times when the Christmas chimes
Were a merry sound to hear,
When the squire's wide hall and the cottage small
Were filled with good English cheer.
Now gold hath the sway we all obey,
And a ruthless king is he;
But he never shall send our ancient friend
To be tossed on the stormy sea.

Then here's to the oak, the brave old oak,
Who stands in his pride alone;
And still flourish he, a hale green tree,
When a hundred years are gone!

The Old Oak Tree

I sit beneath your leaves, old oak,
You mighty one of all the trees;
Within whose hollow trunk a man
Could stable his big horse with ease.

I see your knuckles hard and strong,
But have no fear they'll come to blows;
Your life is long, and mine is short,
But which has known the greater woes?

Thou hast not seen starved women here,
Or man gone mad because ill-fed –
Who stares at stones in city streets,
Mistaking them for hunks of bread.

Thou hast not felt the shivering backs
Of homeless children lying down
And sleeping in the cold, night air –
Like doors and walls in London town.

Knowing thou hast not known such shame,
And only storms have come thy way,
Methinks I could in comfort spend
My summer with thee, day by day.

To lie by day in thy green shade,
And in thy hollow rest at night;
And through the open doorway see
The stars turn over leaves of light.

W. H. Davies (1871–1940)

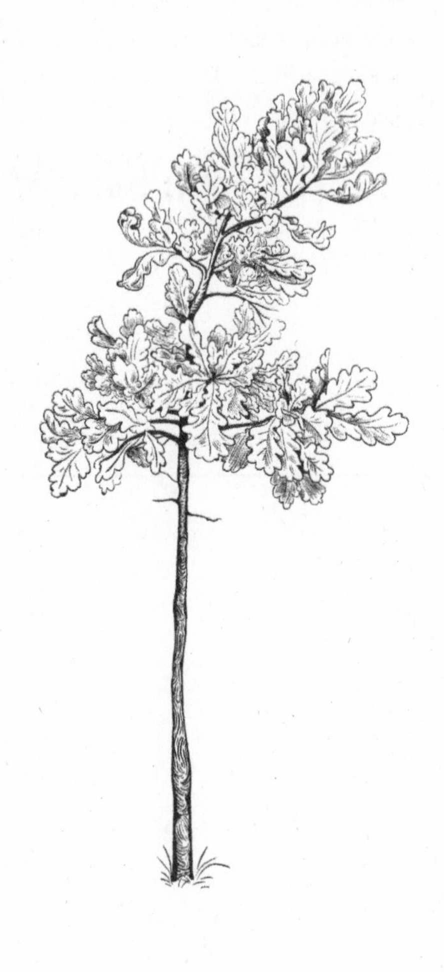

II

Oakwatch: The Life Cycle of the Oak

No tree in all the grove but has its charms,
Through each its hue peculiar; paler some,
And of a wannish gray; the Willow such,
And Poplar, that with silver lines his leaf,
And Ash, far-stretching his umbrageous arm;
Of deeper green the Elm; and deeper still,
Lord of the woods, the long-surviving Oak.

William Cowper (1731–1800)

THE OAK IS Britain's most massive tree – not to be confused with the tallest, a fir – and apart from yew, the longest lived. A mature oak can contain a hundred tons of wood. In botanical jargon it is also a 'keystone plant', supporting more species than any other native tree.

The two British oaks, *robur* and *petraea*, are the most northerly representatives of a vast genus of deciduous trees, *Quercus*, which contains over five hundred species worldwide. While both the British varieties appear across the isles, *Quercus robur*, the penduncluate or common oak, is typical of the deeper, more water-retentive soils of the south and Midlands. *Quercus petraea*, the sessile oak, is characteristic of the rugged hills of the west and north.

As oaks mature they form a broad and spreading crown with sturdy branches beneath. Their open canopy enables light to penetrate through to the woodland floor, allowing bluebells and primroses to, as Robert Bridges versed it, grow below:

Thick on the woodland floor
Gay company shall be,
Primrose and Hyacinth
And frail Anemone

The oak's smooth and silvery brown bark becomes rugged and deeply fissured, like cracks in rock, with age. The fissures in the oak – far deeper and more numerous than in beech or ash – is useful shelter for small creatures, especially arthropods, the spiders, the centipedes, the creatures with many legs.

The rule of thumb for ancient oaks is that they grow for three hundred years, mature for another three hundred years and then 'veteranize', or decay, for another three hundred. John Dryden (1631–1700) put it more poetically:

The monarch oak, the patriarch of the trees,
Shoots rising up, and spreads by slow degrees.
Three centuries he grows, and three he stays
Supreme in state; and in three more decays.

To extend their lifespan, oaks will shorten with age. A 'stag-headed oak' is an oak which has 'retrenched', abandoned its upper branches so they go as bare as bone, in order to reduce the surface area of trunk to be supported by leafage. Shakespeare described the phenomenon in *As You Like It*:

Under an oak, whose boughs were moss'd with age,
And high top bald with dry antiquity

Although oaks can produce acorns at five years of age, they need to be approaching forty years or more before they make significant amounts of seeds, which are concentrated in mast years, every two to three summers. Most acorns will never get the chance to germinate; they are a rich food source, eaten by many wild creatures including mice, squirrels, pheasants, wild duck (surprisingly), voles, mice, deer, rooks and wood pigeon; a 'woodie' can consume a hundred acorns in a day. Jays are particularly partial to acorns, hence the second part of the birds' scientific name, *glandarius*, meaning exactly that, 'eating acorns'. Jays frequently bury acorns for a rainy food day, and the ones they forget may become new oaks. Jays are nature's own oak-planters.

Despite their high numbers in Britain, oak trees are threatened by a number of pests and pathogens. Oak, though, has weapons: an oak attacked by waves of insects in June can shed an entire complement of leaves and grow another the same summer. Oaks can also deploy chemical trickery: tannin stored in the bark is released as vapour when the tree is under attack from leaf-devouring insects.

Whether the oak can fight back against its most pernicious enemies only time will tell. The oak processionary moth is a non-native pest that has been found in London and Berkshire. Not only does it damage the foliage of the trees and increase the oak's susceptibility to other diseases, it is a risk to human health. The moth's hairs are toxic and can lead to itching and respiratory problems if inhaled.

The decline of mature oaks first aroused concern in the 1920s, and today most cases are in central, southern and eastern England. Key symptoms of so-called acute oak decline (AOD) and chronic oak decline (COD) include canopy thinning, branch dieback, black weeping patches on stems and lesions underlying the bleed spots.

Neither is oak mildew, in which the leaves of oak trees are left covered with a white fungal wash spread by wind-borne spores, the trivial disease it seems. Oaks seem to have mysteriously lost the ability which they had until the nineteenth century to grow easily from seed in existing woods, and oak mildew has been implicated in the barrenness.

The good news? It remains extremely easy to grow an oak from an acorn in a garden.

The Oak in Winter

Winter is the time to first view an oak, when it is bare of leaves and stands in its naked truth. True, one or two specimens will retain their foliage, rust-red, till almost spring, an ancestral quirk. But only one or two.

A British winter usually creeps in; rarely does it slam shut with the iron clang of the prison door. But it is cold, whether gradual or acute, that provokes the abscission layer between leaf and stem, causing the oak leaf to swing-float to the ground. Even in its dying, an oak leaf is a thing of art.

Put your fingers in the rock-cracks of the oak's trunk, and you will feel the spider's habitat. Some adult spiders, such as *Clubiona brevipes*, overwinter in the crevices. For those insects that overwinter as eggs, the twigs as well as the bark are a hibernaculum.

In whatever form insects spend the winter, they are a food source for the birds, who have no 'dead time'. A December oak will swarm with gangs of long-tailed tits and goldcrests. Long-tailed tits, which have close kin-bonds, call incessantly to keep in touch.

An oak in winter attracts birds, as light lures moths, sugar pulls bees. Firecrests, wrens, blue tits, great tits and coal tits all descend on the grey-skied oak. The treecreeper will be there too, with his stiff steadying prop of tail, grasping the trunk, head up, as he inspects the work to be done. Of the standard woodland birds, only the treecreeper has a bill long enough yet fine enough to probe mature oak bark and successfully remove insect eggs or larvae.

At night, the treecreeper will roost in a hole or under the ivy on the oak's trunk; ivy contains a chemical anti-freeze, which protects the cells during hard weather, so it remains unseasonally glossy and bright. Several British bat species may also roost under the ivy, in old woodpecker holes, under loose bark or in hollow branches. As many as a hundred pipistrelles will gather together for the winter sleep. Although quiescent in hibernation they can be detected by their pungent, pissy smell.

In the closing light of a winter's day, the mosses and lichens which clothe the bark of the trunk add a subtle beauty; as many as thirty different lichens inhabit the oak's skin – the raggedness of the bark provides an ideal footing for these epiphytes. The most conspicuous of these strange symbiotic beings, part fungus, part alga, is *Ramalina farinacea*, which resembles grey deer antlers. You may be lucky enough, as I once was, to see a fallow deer buck under an oak branch hanging with *Ramalina farinacea*, so that bone antler mirrored lichen antler. In poor, cold winter deer will eat the oak's bark and twigs, favouring the brown-scaly buds which have already formed, and sit waiting for spring.

The world is old and hard, and under an ancient oak in the bleak solitude of a winter's afternoon you know it. I have never known an owl nest in oak, though they favour it for a look-out. In winter the tawnies screech for a mate, and the harsh sound echoes off the oak's stony limbs.

The people of bygone times thought trees the connecting link between air and earth; look at the silhouette of the winter oak in moonlight and you will see it as an arm and hand thrust up through the earth grasping for air, which in a way it is.

Under the woodland floor, the oak has an unseen life. Subterranean fungi – mycorrhizae – weave into the tips of the tree's roots and extend through the soil to form an underground internet, via which can be sent nutrition (sugar, nitrogen, phosphorus), along with warnings about such crucial arboreal matters as aphid attack. The so-called 'Wood Wide Web'.

The relationship between these mycorrhizal fungi and the trees is old, and mutualistic. The fungi siphon off food from the trees, taking some of the carbon-rich sugar that they produce during photosynthesis. The oaks, in turn, obtain nutrients the fungi have acquired from the soil, by use of enzymes that the trees do not possess. One common fungus to extract food from oak roots is the honey fungus, *Armillaria mellea*.

In winter the oak is in a state of dormancy or diapause, which is a sleep, not a death. The oak ticks over, waiting for the light and heat switches of spring.

The saying 'by hook or by crook' comes from the Middle Ages when villagers were only allowed to take dead wood, not cut down trees or bushes. Fallen timber and dead wood could be cleared and pulled out with a shepherd's crook or a weeding hook. Nothing was more prized than oak. A living oak in winter is cold; the oak on the fire burns slow, and solidly. As good as coal.

The Oak in Spring

In nature, the oak is among the first indicators that spring has warmed the ground, and reawakened the sleeping plants. Just as the willow shows the new season in a yellowing of the twigs, the oak blushes a premature pale green. Curiously, although the colour change in the oak is caused by the swelling of the bud, the phenomenon is best seen by walking away from the tree to view it in its entirety.

A double proof of spring is when the male mistle thrush flies to the top of the oak and sings. A mistle thrush, in the bare choir stalls of an oak, always impresses with his physical bulk. He is Pavarotti in feathers.

During the depths of winter a mistle thrush will sit in his food store, a holly bush or a mistletoe clump (hence the name), guarding it against all comers. With the turning up of the temperature dial, he emerges to stake, through song, a claim to breeding land and wife. So begins the yearly cycle of mating and birdsong.

Whatever the weather, the mistle thrush will chanson, hence his country name of 'storm cock'. He may well be joined by the great tit, who similarly likes a high post to 'ring his bell', as country people used to say. His call is a rusty, rather monotonous *hi-tatty, hi-tatty, hi-tatty*.

The mistle thrush nests early, and often in the fork of an oak. The nests are bulkily obvious in oaks without full leaf. To compensate for conspicuousness, the mistle thrush is aggressive. Trespassing buzzards, foxes, cats, crows will all receive a screaming dive-bombing. Aptly, the Welsh call the mistle thrush *penn-y-llwyn*, meaning 'master of the copse'.

Other birds nest inside holes in the tree. Holes and crevices in the bark are ideal homes for the pied flycatcher or marsh tit. Green woodpeckers excavate their own nest holes in the decaying wood of the oak; the hole is chiselled rather than hacked out, the beak being inserted into a crack then twisted to prise off flakes of wood. The nuthatch, a bird almost as colourful as a kingfisher, will plaster up other holes with mud, of pounds in weight, to make the correct-sized entrance.

Spring. Botanically the ground beneath the oak is now at its most picturesque. Wood sorrel and wood anemone are the first to show. Wood sorrel is light sensitive, and according to old country lore you can tell the time of day by the degree to which the flowers have opened into bells. The fragile wood anemone is also known as 'windflower', and truly does shimmy with the slightest of breezes. It dies almost immediately if picked.

As the days progress, the bluebell flowers create pale mauve pools around the oak.

Plants in woodland have evolved different strategies to cope with the problem of shading by the canopy. Wood anemones, wood sorrel and bluebells flower early before the oak is in full sail. They also produce their leaves now; these soon die away, leaving the bulb to store energy for next year's growth. Other plants flower later, so as to avoid the competition with the early species. An example is yellow archangel.

Up in the oak itself, the buds open still further; proper 'leaf-burst' comes mid-May. The full-grown leaves are long, about 8–12cm with a precise 'lobey' design, as if stamped by pastry cutters. Together with the unfolding green leaves come the oak's flowers.

Oaks are 'monoecious', having both male and female flowers on the same tree. The male flowers are catkin-like pale tassels; the female flowers, and it is these which will ultimately form acorns when fertilized, are discreet short spikes. All that protrudes to catch the wind-blown pollen are three tiny reddish styles.

The flowers and leaf buds of the oak are foodstuff for many. Bullfinches gorge on the male flowers, as do caterpillars of purple hairstreak butterflies and the winter moth. Extraordinarily, the caterpillar of the small moth *Dystebenna stephensi* eats the living bark of the tree.

On light spring winds the migrant birds come in, led by the pathfinding chiffchaff, who returns to his familiar oak, to sing his song of only two tones: *chiff*; then *chaff*. It is a wonder how this scrap of feather can manage to fly a yard, let alone thousands of miles from Africa.

By late April, the migrant birds come all in a rush: blackcap, willow warbler, whitethroat, lesser whitethroat, and in the sessile woods of the west, the redstart, the tree pipit and the pied flycatcher. Last of all to arrive is the cuckoo.

No herald of spring is the cuckoo, despite all the excited letters to *The Times* over the centuries. The anonymous thirteenth-century poet had it right when he wrote:

Sumer is icumen in. Lhude sing cuccu!

The Oak in Summer

By now, the oak has erected its high green roof. The clotty-clumps of leaves, however, do let some light to the woodland floor, unlike the beech, whose base is in near-permanent darkness. In the shadow and the shade a favourite of oak woods blossoms; this is the sun-yellow perforate St John's Wort, used by herbalists as a treatment for depression. Beside it, in the leaf litter the woodcock scrapes its hide-and-seek nest, while the rufous plumage of the sitting female blends her to the floor. The woodcock is best, if unkindly, described as 'dumpy'. The male is famous for its 'roding' display flight, the female legendary for carrying her young between her legs. Woodcock are crepuscular, coming out like ghosts to feed in the evening, when they probe the woodland floor with long bills with a super-sensitive tip that allows them to feel for subterranean earthworms and insect larvae.

The real life of the summer oak is up in the tree's canopy, where the air huzzes with insect life. Shaking an oak branch on to a white sheet gives an idea of the multitude and the numbers and species involved. No fewer than a hundred different moth types have been recorded on oak, one of the commonest being the green oak tortrix, whose caterpillar is a velvety pale green and just under an inch long; they are 'loopers', moving by humping their back and bringing their tail forward in a jerky motion. If they fall from a leaf they spin a silk 'lifeline' that enables them to climb back up and continue feeding. The oak beauty moth's caterpillars also enjoy oak foliage. When caterpillar numbers are high, they can be heard on still afternoons eating away at the oak's leaves, and their droppings or 'frass' fall like sticky black rain.

The grubs of some moths, sawflies and flies 'mine' inside the oak's leaves. The crop of caterpillars and grubs is of vital importance to the tits and warblers of the oak wood. By some unfathomable design of nature, the small perching birds correlate the birth of their chicks with the hatch of insect larvae.

Moth caterpillars can strip an oak of its primary leaves. If this happens, the tree has a second spate of leaf growth, so-called 'Lammas shoots' that burst forth in the late summer, in time to photosynthesize and produce food for winter.

Strangely, while the oak supports so many moth species, few butterflies are associated with it. The most elegant of the oak butterflies is the purple emperor, so named because the male's wings sheen with the colour traditionally worn by royalty. (Queen Elizabeth I forbad anyone except close members of the royal family to wear it.) The larvae of the purple emperor feed on sallow, but when adults they float through the tops of oaks, their main habitat in July and August. The food they seek on oak leaves is honeydew, the sugar-rich excrement of aphids.

As summer progresses the oak's leaves build up higher concentrations of tannin. This makes them less palatable to larvae, and dark to the eye. Long gone is the springtime translucence, when every oak leaf was a pane of green glass. Look carefully at the leaves of late summer and you will spot small red buttons on the underside. These are 'spangle-galls', a sort of abscess produced by the tree in response to eggs laid by gall wasps, in this case members of the genus *Neuroterus*.

It is during summer's days that the leafy bract at the base of each female flower turns woody, and fuses to form a hard green cup. A single seed develops in this cup. So do acorns grow. The female acorn weevil, an improbable long-nosed character more suited to a Disney cartoon than tooth-and-claw real nature, bores a small hole inside a developing acorn in which she lays a single egg. The resultant larva lives inside the acorn, eating its flesh, until autumn when the ripe acorn falls to the ground, and the fat adolescent larva crawls out to pupate in the soil.

If our oak hosts dedicated and exclusive species, it also harbours opportunists. The woodcock is a wader which has roamed far from the shore. A weasel can climb an oak, and into a green woodpecker's nest to eat the chicks; it may gorge so extensively that it is unable to exit until it has slept and thinned. Once, I saw a grass snake slither up a tree, coiling around every lump and bump for purchase, to take the egg of a naively low-nesting wood pigeon.

By July birdsong in the oak will have died down. Most young are on the wing and the hard work of rearing is over, though some birds will have second or even third broods. One reason for the silence in the oak is the business of the moult, which reduces the power of flight. No bird wants to advertise incapacity. Still, some will fall prey to raptors. The ecology of the oak tree is a game of consequences: hairstreak caterpillars feed on the tree's leaves; the coal tit feeds on the crawling hairstreak caterpillars; the sparrowhawk flashes through the branches, a cutting blade, to catch and feed on the coal tit. The male sparrowhawk is smaller and nimbler than his mate, and is the terrorist of the canopy.

The well of shade of the oak in the field is welcome for the sheep and cattle. Sometimes under the green shade of the oak in the wood, the deer will stand quivering, trying to escape the heat and the flies. They seem to await a royal hunter, or Robin Hood, in a scene from another, now long gone, Britain.

Under their hooves, the roots of the summer oak will draw up 50 gallons or more of water per day.

The Oak in Autumn

According to John Keats, autumn is the 'season of mists and mellow fruitfulness'. For oaks in the west and north of Britain, autumn is equally likely to be the water-time. Black oaks in the rain are columns holding down the wind-ripped land.

In autumn the oak prepares its survival system for winter. The sappy nutrition the leaves have produced through photosynthesis during the summer is transferred through the oak's vascular system – akin to the veins and arteries of a human – to the roots. Leaves, which are tender and vulnerable to frost, are shed.

In a screaming Halloween wind, the leaves are sometimes gone in a night, like revellers from a city centre at 3am. Left behind, high up in the trees, are the blots of the wood pigeon's nest and the carrion crow. That night the leafless oak will hold the stars in its branches.

The sodden leaves of oak break down with ease in autumn and form a rich leaf mould beneath the tree, which catches in the nose like churchy incense and supports such invertebrates as the stag beetle, and such fungi as the oakbug milkcap. Larger animals, such as shrews and mice, rely on the leaf

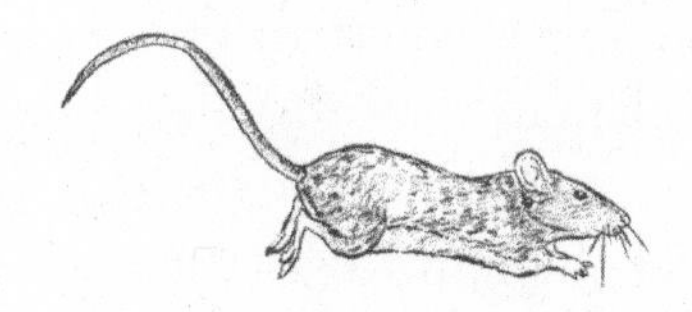

litter for their survival in the cold; snow and frost will not penetrate the 'carpet', allowing them to carry on their scuttling lives unhindered.

As the leaves fall, 'oak-apple galls' become easier to spot. These are the most conspicuous result of an insect attack on the oak, a small, round, light brown protuberance on the twigs. The gall is caused by a tiny wasp, *Biorhiza pallida*; the wasp lays an egg in the twig, causing the oak to produce a chemical that forms a protective structure around the egg. Other common galls on the oak are the 'knopper gall', caused by *Andricus quercus*, and the 'marble gall' produced by the wasp *Andricus kollari*. Marble galls contain a concentration of 17 per cent tannin. The wasp was deliberately introduced to Britain in around 1840 to assist the leather-tanning industry.

The wind brings down the acorns, which are green initially, but if allowed by the elements to ripen in the cup they go nut-brown.

As the oak grows older and stouter, it will be invaded by mistletoe and ivy, and in wet districts by mosses, lichen, algae and ferns. The wind carries the spores of the common polypody fern to the bark of the old tree, where enough vegetable matter has collected in the pits and pores of age for the fern to grow as readily as in soil.

Eventually our oak will be invaded by one of the many fungi that cause the decay of its heartwood, and eventual downfall. The sulphur bracket fungus develops from a spore that enters an open wound, such as the open socket where the wind has broken off a branch; the fungal threads then penetrate into the oak's heartwood, until the trunk breaks under the weight of its crown or from the force of the ever-returning winds.

Autumn is the rutting season for deer. The bucks bark with testosteroned excitement beneath the oak. Although the robin prefers the gatepost, he or she may sing in the lower branches of the oak – sharp, reflective phrases, as if philosophizing on the life and times of our familiar tree.

Oak Oddities

- Quercivorous means 'feeding on oak leaves'.
- King Arthur's round table is believed to have been made from a single piece of ancient oak.
- In the 1820s 90,000 tons of oak bark were used per annum in Britain by the tanning industry. Oak bark, stripped from felled trees, is still valued at several hundred pounds per ton for tanning leather.
- 'Tannin' is derived from *tanna*, Old High German for oak.
- The Oaks Estate near Epsom, Surrey, gave its name to the famous annual flat race for three-year-old fillies, 'The Oaks'.
- A shillelagh, taking its name from a village in County Wicklow, is an Irish cudgel of oak or blackthorn.
- 'Oak' is Cockney rhyming slang for 'broke', meaning penniless. 'Ash' is 'cash'.
- Due to its close, hard grain oak was used in medieval times for water pipes and buckets.

- Oak has low electrical resistance, and was chosen for the construction of the first electric chair, used in New York State in 1890.

- There are three introduced species of oak in Britain, the American red, the Mediterranean holm (an evergreen with the appearance of a giant holly bush), and the Turkey, which has distinctive zigzag lobes.

- Located in the French farming village of Allouville-Bellefosse, in Seine-Maritime, is the Chêne Chapelle, a hollow, thousand-year-old oak trunk which serves as the base for two small chapels accessible via spiral staircases that surround the tree.

- 'Bog oak' is oak that has lain for centuries in an acid peat bog; the dark timber is prized for ornamental woodwork.
- Gypsy children are buried with an acorn in each hand.

III

'To Assuage Inflammations': The Oak as Medicine, Food and Drink

To the British, the oak was as the buffalo to the Sioux. The all-provider. After praising the oak as the 'glory and safety of this nation by sea', the Elizabethan herbalist Nicholas Culpeper (1616–1654) settled down to listing the tree's medicinal 'virtues':

> The leaves and bark of the Oak, and the acorn cups, do bind and dry very much. The inner bark of the tree, and the thin skin that covers the acorn, are most used to stay the spitting of blood, and the bloody-flux. The decoction of that bark, and the powder of the cups, do stay vomitings, spitting of blood, bleeding

> at the mouth, or other fluxes of blood, in men or women; lasks [diarrhoea] also, and the nocturnal involuntary flux of men. The acorn in powder taken in wine, provokes urine, and resists the poison of venomous creatures. The decoction of acorns and the bark made in milk and taken, resists the force of poisonous herbs and medicines, as also the virulency of cantharides [Spanish fly], when one by eating them hath his bladder exulcerated, and voids bloody urine. Hippocrates saith, he used the fumes of Oak leaves to women that were troubled with the strangling of the mother; and Galen applied them, being bruised, to cure green wounds. The distilled water of the Oaken bud, before they break out into leaves is good to be used either inwardly or outwardly, to assuage inflammations, and to stop all manner of fluxes in man or woman. The same is singularly good in pestilential and hot burning fevers; for it resists the force of the infection, and allays the heat. It cools the heat of the liver, breaking the stone in the kidneys, and stays women's courses.

Inexplicably, he forgot to mention that village matchmakers insisted on the efficacy of May dew gathered from oak leaves as a beauty treatment for young women, that mistletoe from oak was particularly valued in treating 'falling sickness', known today as epilepsy, that fingernail clippings and leg hairs stuffed in an oak hole bunged up with cow dung was said to get rid of gout.

And that to stay healthy for a whole year, the Welsh believed you should run your hand over an oak branch, that a nail driven into the gum above a hurting tooth then driven into an oak transferred the pain. During the seventeenth century the oak was considered a remedy for baldness (by washing the head in water used to soak leaves and 'midle rinde' of an oak), and if an oak branch was cut and stepped over then the subject was protected against witchcraft. According to the *Philosophical Transactions of the Royal Society*, 1672, limbs could be strengthened 'by anointing them with Oyl drawn out of the white Oak acorns'.

As food the oak's fruit was most valued in times of austerity. In the second century AD the Roman doctor Galen recorded poor country folk making flour from acorns. Geoffrey Chaucer mentions some who 'were wonte lightlie to slaken hus hunger at even with akehornes of okes'. Acorns, thick with tannins, are inedible raw. To remove the bitterness, Native Americans buried acorns in the earth and let it leach out. Boiling achieves the same result. Rural Koreans make *dotorimuk*, acorn jelly.

In Germany in the Second World War, acorns were roasted and ground as an ersatz coffee.

The oak's leaves and wood still have a place on the table. Springtime oak leaves make a light white wine, and the tree's wood traditionally smokes herrings in Caister and haddock in Arbroath, while oak casks flavour and mature brandy, beer, wine and spirits.

Acorn Coffee

The recipe for 'acorn coffee' is simple enough, but by no means could the drink be said to taste like anything made from Brazilian beans: on the tongue acorn coffee is reminiscent of the Barleycup that was once a staple of health food cafés.

Peel the acorns, boil for ten minutes, leave to dry for a day. Then roast on the middle shelf of the oven at 120 degrees C for about 15 minutes, before grinding in a coffee grinder. Put the ground acorn powder into a cafetière or percolator, at the rate of 2 teaspoons per cup, and add boiling water. Drink.

Giliø kava is a Lithuanian traditional drink, and a variation of acorn coffee, whereby the shelled, dried acorns are gently simmered in milk until soft (about an hour), then strained, patted dry, and scorched in a hot frying pan. The acorns are then ground, and treated like 'coffee'.

Oak Leaf Wine

Ingredients:

4.6 litres (1 gallon) of fresh young leaves
4.6 litres (1 gallon) of water
900g (2lb) sugar
juice and zest of 2 or 3 oranges
1 teaspoon yeast

Method:

1. Pour boiling water over the leaves, stir well and leave overnight.
2. Strain off the leaves and bring the liquid to the boil. Add the squeezed orange juice and finely grated rind.
3. Allow the liquid to become lukewarm, add the yeast and pour into a demijohn. Fit a fermentation lock and add water to the top of the lock. Rack at one month, and two months.
4. When fermentation ceases, siphon the liquid from the sediment and bottle. The wine will be ready to drink after a year.

From 'Yardley Oak'

Survivor sole, and hardly such, of all
That once lived here, thy brethren, at my birth
(Since which I number threescore winters past)
A shattered veteran, hollow-trunked perhaps
As now, and with excoriate forks deform,
Relicts of ages! Could a mind imbued
With truth from Heaven created thing adore,
I might with reverence kneel and worship thee.
It seems idolatry with some excuse
When our forefather druids in their oaks
Imagined sanctity. The conscience yet
Unpurified by an authentic act
Of amnesty, the meed of blood divine,
Loved not the light, but gloomy into gloom
Of thickest shades, like Adam after taste
Of fruit proscribed, as to a refuge, fled.
Thou wast a bauble once; a cup and ball,
Which babes might play with; and the thievish jay
Seeking her food, with ease might have purloined
The auburn nut that held thee, swallowing down
Thy yet close-folded latitude of boughs
And all thine embryo vastness, at a gulp.
But fate thy growth decreed. Autumnal rains

Beneath thy parent tree mellowed the soil
Designed thy cradle, and a skipping deer
With pointed hoof dibbling the glebe, prepared
The soft receptacle in which secure
Thy rudiments should sleep the winter through.
So Fancy dreams. Disprove it if ye can
Ye reasoners broad awake, whose busy search
Of argument, employed too oft amiss,
Sifts half the pleasures of short life away.
Thou fell'st mature, and in the loamy clod
Swelling, with vegetative force instinct
Didst burst thine egg, as theirs the fabled twins
Now stars, two lobes protruding paired exact.
A leaf succeeded, and another leaf,
And all the elements thy puny growth
Fostering propitious, thou becamest a twig.
Who lived when thou wast such? Oh could'st thou speak [...]

William Cowper (1731–1800)

IV

Britain's Mightiest Oaks

Of all the trees that grow so fair,
Old England to adorn,
Greater are none beneath the Sun
Than Oak and Ash and Thorn.

Rudyard Kipling (1865–1936)

THERE ARE OVER 700 named oaks in Britain. Visitors will find these veteran oaks in various moods; an oak has many looks, dependent on the season. Mature oaks are hard to miss. They can be 100 feet tall, with a spread of 150 feet across, and a girth of more than 30 feet.

Haresfield Oak, Gloucestershire

With a girth of more than 23 feet, this oak was reputedly planted to mark the route that the funeral cortège for Edward II took from Berkeley Castle to Gloucester in 1327.

Druid's Oak, near Farnham Common, Buckinghamshire

Has a girth of nearly 30 feet and is thought to be up to 1000 years old. Felix Mendelssohn is said to have composed some of the music for *A Midsummer Night's Dream* while visiting the surrounding woods.

Dickens Oak, Chigwell, Essex

At least 300 years old, with a girth of almost 20 feet, the tree is near to the King's Head public house, portrayed by Charles Dickens as the Maypole in *Barnaby Rudge*.

Law Day Oak, Bonnington, Kent

Used as a venue for the administration of justice since at least the reign of Elizabeth I. An annual parish meeting is still held beneath its branches, the last vestige of the moots once held up and down England.

Bowthorpe Oak, Lincolnshire

One of Britain's oldest oaks, estimated at over 1000 years old, already regal when the peasant poet John Clare celebrated it in verse in the 1820s:

Old noted oak! I saw thee in a mood
Of vague indifference; and yet with me
Thy memory, like the fate, hath lingering stood
For years, thou hermit, in the lonely sea
Of grass that waves around thee! Solitude
Paints not a lonelier picture to the view,
Burthorp! than thy one melancholy tree
Age-rent, and shattered to a stump. Yet new
Leaves come upon each rift and broken limb
With every spring; and Poesy's visions swim
Around it, of old days and chivalry;
And desolate fancies bid the eyes grow dim
With feelings, that Earth's grandeur should decay,
And all its olden memories pass away.

According to Bartholomew Howlett, in *A Selection of Views of the County of Lincolnshire* (1805), one owner of the tree, a certain George Pauncefort, Esq., 'had the interior of the oak floored, with benches placed round and a door of entrance where 12 persons had frequently dined in it with ease'. Another former owner used it as a place to feed his calves, a not uncommon use for a hollow oak. (An oak so purposed was called a 'bull oak'.) Another owner erected a pigeon house in its canopy. The tree is located on Bowthorpe Park Farm, off the A6121, and visitors are welcome throughout the year. The girth is 42 feet, the largest in the UK.

Queen Elizabeth Oak, Northiam, East Sussex

On 11 August 1573, Elizabeth I stopped in the village on her way to Rye and sat beneath this spreading oak to eat a meal served to her from the house nearby, owned by George Bishop and his family. She changed her shoes of green damask silk and left them as a souvenir. In 1944, troops paraded beside the tree prior to D-Day. The original oak died in the late twentieth century, but a sapling stands in its place.

Eardisley Oak, Lower Welson, Herefordshire

Easily found, since it has its own brown heritage sign. Also known as the Great Oak, the tree has a massive girth, 31 feet, indicating a lifespan of about 900 years.

As with all antique oaks, it is 'stag-headed'. A forest in this area was recorded in the Domesday Book, and this may be its last surviving tree. The Eardisley oak was a landmark on maps dated 1650. Herefordshire is the top county for ancient oaks, with 366 oaks older than 400 years.

Jack of Kent's Oak, Kentchurch Court, Herefordshire

A stag-headed oak with a girth of 38 feet or so. Jack o' Kent or Jack-a-Kent was a folkloric character of the Welsh Marches, sometimes described as a wizard, sometimes as a priest. His modus operandi was to take on the devil in bets and games, and beat him. One story tells how he asked the devil to help him build a bridge over the Monnow river; in return the devil could have the soul of the first to cross the bridge. The bridge-building accomplished, Jack threw a bone across the bridge and a hungry dog galloped after it, so sparing a human life.

King of Limbs, Savernake Forest, Wiltshire

As the name suggests, this tree's branches and roots sprout out, like legs and arms. The Savernake Forest has been dated back over 1000 years to before the time of William the Conqueror, and legend has it Henry VIII courted Jane Seymour beneath its great oaks. (Her father, Sir John, was the forest steward.) Thought to be 1000 years old, the King of Limbs is a pollard, and the inspiration for Radiohead's 2011 album of that title. Nearby, almost on the verge of the A364, is Belly Oak, a hybrid *Quercus x rosacea*, and another possible remnant of the original historic forest.

Major Oak, Nottinghamshire

Probably the most famous, most visited, tree in the whole of the UK, the Major Oak's hollow trunk was supposedly used as a hideout by Robin Hood's merry men. In a mast year it has produced an estimated 150,000 acorns. Age has wearied this pollard, which is supported by posts. Named after Major Hayman Rooke, author of *Remarkable Oaks in the Park of Welbeck*, 1790.

Queen Elizabeth Oak, Hatfield House, Hertfordshire

This is the tree Elizabeth I reportedly sat under when she heard the news of her sister Mary's death, and therefore her ascension to the throne.

It appears that Gloriana had several encounters with oak trees, aside from the oak at Northiam (qv). At the Queen's Oak in Huntingfield, Suffolk, she shot a buck in 1558; the Kiss Oak in Gorhambury Park, Hertfordshire, gained its name because under its green shade she was said to have been caught kissing the Earl of Leicester. The Queen Elizabeth Oak, Greenwich Park, London, was a favourite picnic spot, though the tree's association with royalty goes back to her father, who is said to have danced around it with Anne Boleyn. The tree was brought down in a heavy storm in 1991 but it remains, horizontal and slowly decaying, in the park with a young oak planted beside it.

Royal Oak, Shropshire/Staffordshire border

In 1651 Charles II hid from parliamentarians in an oak tree in Boscobel Wood. To celebrate the King's restoration in 1660, Royal Oak Day was made into a countrywide celebration on 29 May, the monarch's birthday. The original Royal Oak at Boscobel was destroyed in the eighteenth century by tourists snipping off branches as souvenirs. Strictly speaking, the Royal Oak that remains is an offspring of the original but it is still about 300 years old.

Birnam Oak, Perth and Kinross

This ancient oak tree is said to be one of the last trees left of old Birnam Wood, the great forest famously referred to in Shakespeare's *Macbeth*, when the witches foresaw that 'Macbeth shall never vanquish'd be until/Great Birnam Wood to high Dunsinane Hill/shall come against him'. When Malcolm's troops use branches from the wood to disguise their advance on Macbeth's castle, the prophecy comes true.

With its half-hollow trunk and crutches, the appearance of this tree is medieval. J. R. R. Tolkien, taken with 'bitter disappointment' over Shakespeare's shabby trick with Birnam Wood in *Macbeth*, longed to devise trees that really could 'march to war'. Hence the Ents.

Old Knobbley, Mistley, Essex

A pollard oak, situated in the middle of Furze Hills wood, Old Knobbley, with its girth of over 30 feet, looks curiously akin to a portly giant rising from the earth. The tree dates from the thirteenth century. In 2013 it was the runner-up in the Woodland Trust's 'Tree of the Year' competition.

Wyndham's Oak, Dorset

Named for Sir Hugh Wyndham, appointed a judge by Richard Cromwell during his short protectorate. The tree has been used as a gibbet, or hanging tree.

Bradgate Park, Leicestershire

A row of ancient pollarded oaks, allegedly 'decapitated' in 1554 by foresters as a sign of respect, following the beheading of Lady Jane Grey who was born at nearby Bradgate Hall.

Gog and Magog, Glastonbury, Somerset

At the foot of Glastonbury Tor stand Gog and Magog, two very ancient oaks. It is thought they may be the last remnants of an avenue of oaks leading up to the Tor. The origin of the names is obscure. In his *Historia Regum Britanniae* the twelfth-century monk Geoffrey of Monmouth tells of the giant Goemagot, and the theory goes that the name has been divided. Alternatively, some say that the names are derived from the Celtic deity Ogmios and his consort Magog.

Oak Proverbs

The oak is recorded in many British proverbs, such as:

An oak is not felled in one stroke – signifying patience.

Great oaks from little acorns grow – from little things grow great things.

The willow will buy a horse before the oak will buy the saddle – referring to time, as oaks grow much more slowly than willow.

One folkloric belief about weather forecasting is predicated on which tree foliates first:

Oak before ash,
In for a splash.
Ash before oak,
In for a soak.

Another bit of old verse is about safety during a thunderstorm:

Beware of the oak, as it draws the stroke,
And avoid the ash as it counts the flash.
Best creep under the thorn, as it will keep you from harm.

'When the oak is felled the whole forest echoes with its fall, but a hundred acorns are sown in silence by an unnoticed breeze.'

Thomas Carlyle (1795–1881)

V

The Thunder Tree: Oak in Folklore, Myth and Legend

As befits a tree with a long life, the oak has long been held as sacrosanct. In the Old Testament the oak was sacred to believers and non-believers alike. 'Then shall ye know that I am the LORD,' it is written in Ezekiel 6:13, 'when their slain men shall be among their idols round about their altars, upon every high hill, in all the tops of the mountains, and under every green tree, and under every thick oak.'

There is no doubt about oak worship in the Classical world. In Greek mythology oaks were dedicated to Zeus, for an oak was said to have sheltered him at birth in Arcadia. Said to be the most ancient oracle of the Greeks, Dodona, a sanctuary of Zeus, was centred around a sacred oak tree. Dryads, the attendants of the

goddess Artemis (Diana in Roman mythology), were the nymphs of oak trees. Any mortal who harmed an oak without first offering propitiation to the tree nymphs was punished by the gods. This was the fate of Erysichthon who, in need of timber, started cutting down a grove belonging to Demeter, a goddess of fertility. As punishment he was condemned to suffer insatiable hunger. On the orders of the Roman emperor Theodosius I (AD 347–95), a Christian, the oak at Dodona was destroyed in 391. Modern archaeologists have planted a new oak on the site. Herodotus recounts how the holy oaks in the forest of Epirus had the gift of prophecy and spoke with a human voice. Even when oaks were felled to build the ship *Argo*, the beams and mast spoke to warn the Argonauts of approaching calamity.

The obvious attributes of the oak, longevity and strength, were widely admired by the earliest western civilizations. Kings wore crowns of oak leaves, as symbol of the god they represented on earth. In Rome the oak was considered the emblem of hospitality, and also of greatness: oak leaves composed the civic crown for men honoured by the city, as Shakespeare describes in *Coriolanus*:

He prov'd best man i' th' field, and for his meed
Was brow-bound with the oak.

Oak leaves were also a battle honour granted to Roman commanders, and have continued to be decorative icons of military prowess. Insignia with oak leaves marked various Nazi honours. The Hitler Youth badge featured oak leaves. The Totenkopfring ('Skull Ring') worn by SS officers had oak leaves topped off with a death's head etched into it.

The Nazis harked back to Teutonic and Scandinavian tree worship, where the oak was the subject of a cult. Thor was said to ride through the sky in an oak chariot. The belief that Thor sheltered under an oak tree during a storm gave rise to the household practice of keeping an acorn on the windowsill or carved on a banister to ward off lightning. Any oak felled by lightning was doubly lucky, and people swarmed over it collecting charms.

Probably, the oak was the thunder god's tree because it is often the tallest tree in the landscape, and thus more often struck by lightning. From this belief it was a short step to perceiving the oak as sacred, the fire-maker.

The tree's longest association with the myth and religion of Britain begins with the Druids, who considered oaks to have mystical powers. Legend has it that oak trees carry the souls of men who have passed away. (Even today people touch wood; a relic from the days when people thought guardian spirits resided in trees.) Druids frequently practised their rituals and worshipped in oak groves and especially cherished the mistletoe that frequents oak tree branches, as Pliny the Elder (AD 23–79) recorded in his *Natural History*:

Upon this occasion we must not omit to mention the admiration that is lavished upon this plant by the Gauls. The Druids – for that is the name they give to their magicians – held nothing more sacred than the mistletoe and the tree that bears it, supposing always that tree to be the robur. Of itself the robur is selected by them to form whole groves, and they perform none of their religious rites without employing branches of it; so much so, that it is very probable that the priests themselves may have received their name from the Greek name for that tree. In fact, it is the notion with them that everything that grows on it has been sent immediately from heaven, and that the mistletoe upon it is a proof that the tree has been selected by God himself as an object of his especial favour.

The mistletoe, however, is but rarely found upon the robur; and when found, is gathered with rites replete with religious awe. This is done more particularly on the fifth day of the moon, the day which is the beginning of their months and years, as also of their ages, which, with them, are but thirty years. This day they select because the moon, though not yet in the middle of her course, has already considerable power and influence; and they call her by a name which signifies, in their language, the all-healing. Having made all due preparation for the sacrifice

> and a banquet beneath the trees, they bring thither two white bulls, the horns of which are bound then for the first time. Clad in a white robe the priest ascends the tree, and cuts the mistletoe with a golden sickle, which is received by others in a white cloak. They then immolate the victims, offering up their prayers that God will render this gift of his propitious to those to whom he has so granted it. It is the belief with them that the mistletoe, taken in drink, will impart fecundity to all animals that are barren, and that it is an antidote for all poisons. Such are the religious feelings which we find entertained towards trifling objects among nearly all nations.

Mistletoe was an obvious fertility aid; when squeezed, the berry juice looks akin to semen.

John Loudon in his *Arboretum et Fruticetum Britannicum* (1838) tells us:

> The druids professed to maintain perpetual fire; and once every year all the fires belonging to the people were extinguished, and relighted from the sacred fire of the druids. This was the origin of the Yule log, with which the Christmas fire, in some parts of the country, was always kindled; a fresh log being thrown on and lighted, but taken off before it was consumed,

and reserved to kindle the Christmas fire of the following year.

The Yule Log was always of oak.

The poet Robert Herrick gives the particulars of lighting the Yule Log in his 'Ceremonies for Christmasse', 1638:

Come, bring with a noise,
My merrie merrie boyes,
The Christmas Log to the firing;
While my good Dame, she
Bids ye all be free;
And drink to your hearts desiring.

With the last yeeres brand
Light the new block, and
For good successe in his spending,
On your Psalteries play,
That sweet luck may
Come while the Log is a-teending.

'Yule' is derived from Yiaoul, the Celtic god of fire, whose festival was at midwinter.

Christianity appropriated the oak, and not just at Christmas. There are holy oaks in English place names – Cressage in Shropshire, which was Cristesache, 'Christ's Oak', in the Domesday Book; or Holy Oakes in Leicestershire. Many parishes used to have a Gospel Oak, a prominent tree, under which part of the gospel was read out during the Beating of the Bounds ceremonies at Rogantide in spring. The custom of beating the bounds accompanied by reading from the scriptures was introduced among Christians about AD 800 by Avitus, bishop of Vienne, France.

The 'Green Man', the primitive's tree spirit often pictured with wise eyes peering from a face composed of oak leaves, can be seen in most old English churches. He is also known as 'Jack-o'-the-woods'. He was satyric, as well as wise. Christianity struggled to suppress the licentiousness connected with pagan oak worship.

The acorn, from which the mighty oak grew, was an obvious fertility symbol. In medieval England, whether or not lovers would marry could be divined by dropping two acorns, one for each of the couple, into a basin of water; floating close together indicated a wedding. Even the Puritan Oliver Cromwell allowed couples to be married under ancient oaks.

After the Restoration, the long-standing May fertility rite of cutting green oak boughs hybridized with Royal Oak Day, which became renowned for bawdiness. Boys armed with sprays of nettles whipped the bare legs of girls not wearing a token of loyalty to the monarch, usually an oak leaf or acorn. In Exeter, 29 May simply became 'Lawless Day'. As late as 1882 the Reverend Edward Bradley of Stretton, in Rutland, observed the postman whipping a maid who had omitted to dress the door with oak leaves.

In areas still more rural than Rutland the merged message of jolly May fertility and monarchism is alive and well. Fownhope in Herefordshire is the setting for an annual walk held on a Saturday near Oak Apple Day by the village friendly society, Hearts of Oak (established 1876). Attendees are warned to 'expect Morris dancing, music and merriment!'

Indeed, they should. My father was a member of the society. So I know how good those village days can be.

Timeline

c.7000 BC – The oak is established in post-Ice Age Britain.

c.2000 BC – Wooden 'henges' are built, prehistoric sacred sites, similar to Stonehenge.

First millennium BC – The shrine of Dodona established in Greece.

500 BC – Iron Age, and half of Britain is already 'de-wooded'.

AD 43 – the Romans build the first bridge across the Medway, using oak for piles.

First century AD – Pliny the Elder reports on the oak cult of the Druids.

Fifth century AD – Anglo-Saxon migration to Britain, undertaken via oaken longboats, begins.

c. 600 AD – The Sutton Hoo ship burial.

AD 970 – Britain's current longest-living oak begins life.

1066 – Norman Conquest, which begins the widespread introduction of deer parks in England and Wales.

1096 – The building of Norwich Cathedral begins.

1558–1603 – The reign of Elizabeth I, who sanctions the first deliberate oak plantation in order to satisfy the national need for oak timber.

1651 – The future Charles II hides in Boscobel Oak after losing the battle of Worcester.

1805 – The 'wooden walls' of Nelson's navy defeat the combined Franco-Spanish fleet at Trafalgar.

1820s – 90,000 tons of bark used per annum in the British tanning industry.

1935 – The National Trust adopts a sprig of oak as its symbol.

Picture Credits

Alamy Stock Photo: World History Archive/Alamy Stock Photo: 9; Authentic-Originals/Alamy Stock Photo: 50; Old Images/Alamy Stock Photo: 55; Chronicle/Alamy Stock Photo: 64; INTERFOTO/ Alamy Stock Photo: 78; **Illustrations by Beci Kelly:** i, 14, 19, 38, 51, 76; **Bridgeman Images:** Salisbury Museum/Bridgeman Images: 80; **Getty Images:** The Print Collector/Getty Images: 16, 20; **Shutterstock.com:** Marilyn Barbone/Shutterstock.com: vii; Morphart Creation/Shutterstock.com: 6, 46; Chempina/ Shutterstock.com: 32; Bodor Tivadar/Shutterstock.com: 53; Concluseratimage/Shutterstock.com: 86.